How Do Things Move?

Fast and Slow

Sue Barraclough

Chicago, Illinois

a division of Reed Elsevier Inc.
Chicago, Illinois

Customer Service 888–454–2279

Visit our website at www.heinemannlibrary.com

Photo research by Ruth Blair, Erica Newbery, and Kay Altwegg
Designed by Jo Hinton-Malivoire and bigtop design ltd
Printed and bound in China by South China Printing Company
10 09 08 07 06
10 9 8 7 6 5 4 3 2 1

Library of Congress Cataloging-in-Publication Data
Barraclough, Sue.
Fast and slow / Sue Barraclough.
p. cm. -- (How do things move?)
Includes bibliographical references and index.
ISBN 1-4109-2261-8 (library binding-hardcover) -- ISBN 1-4109-2266-9 (pbk.)
1. Motion--Juvenile literature. 2. Speed--Juvenile literature. I. Title II. Series: Barraclough, Sue. How do things move?
QC133.5.B37 2006
531'.112--dc22
2005029629

Acknowledgments
The author and publisher are grateful to the following for permission to reproduce copyright material: Alamy pp. **6, 7** (Peter Steiner), **10** (Dennis MacDonald), **19** (David Hoffman Photo Library); Corbis pp. **8, 23 top right**; Corbis pp. **4, 5, 22 top** (Michael Kim), **9, 23 bottom right** (Lester Lefkowitz), **11, 23 top left** (Danny Lehman), **12** (Don Mason), **14, 23 bottom left** (Tom Brakefield), **18** (Randy Faris), **20, 21** (Laura Doss); Digital Vision pp. **15, 22 bottom**; Harcourt Education p. **13** (Tudor Photograhy); NHPA pp. **16** (David Middleton), **17** (Ernie Janes).

Cover photograph reproduced with permission of Stockbyte.

Some words are shown in bold, **like this**. You can find out what they mean by looking in the glossary.

Contents

Fast

This is a race car.

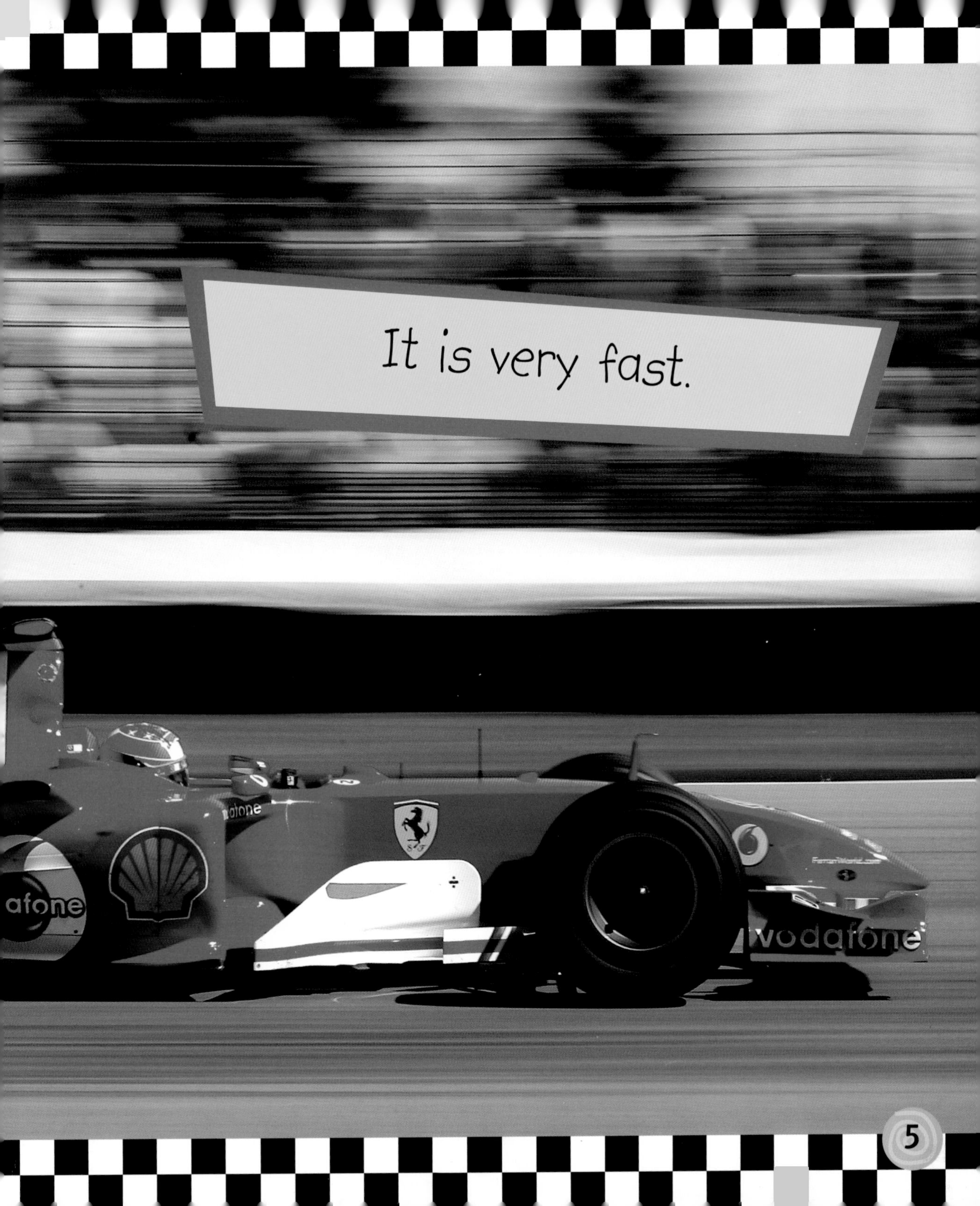

It is very fast.

Slow

This is a **steamroller**.

It is slow.

Soccer
Festival
GALION
NEW YORK
837161

Fast and Slow

This is a rocket blasting into space.

Is it fast or slow?

This is a tractor on a farm.

Is it fast or slow?

Boats and Ships

This is a speedboat on a lake.

This is a big ship on the sea.

The speedboat is faster than the ship.

Motorcycles and Bicycles

This motorcycle has an **engine**.

This bicycle has no engine.

Which is faster, the motorcycle or the bicycle?

Cheetahs and Turtles

This is a cheetah.
It runs very fast.

This is a turtle.

It walks very slowly.

Horses

This horse is running. Is it moving fast or slow?

This horse is walking. Is it moving fast or slow?

Running and Walking

These women are running fast.

This woman is running the fastest.

These people are walking slowly.
This girl is walking the slowest.

Racing at the Beach

These children are racing each other at the beach.

Who is moving the fastest?

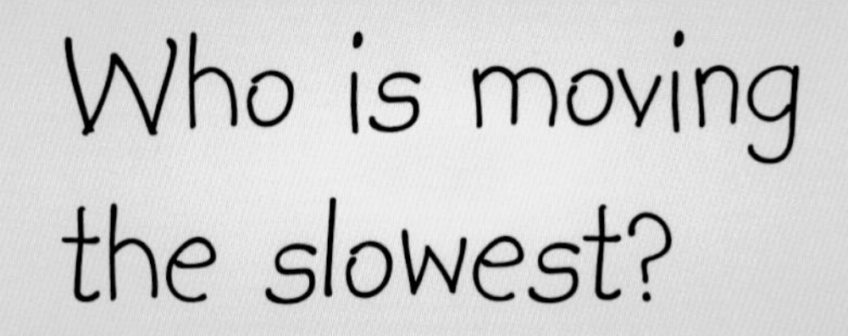

Who is moving
the slowest?

Fast or Slow?

Can you remember what was fast and what was slow?

Glossary

engine machine that makes something move

steamroller heavy machine that makes new roads flat

Index

Notes for Adults

The *How Do Things Move?* series provides young children with a first opportunity to learn about motion. Each book encourages children to notice and ask questions about the types of movement they see around them.

These books will also help children extend their vocabulary, as they will hear some new words. Since words are used in context in the book this should enable young children to gradually incorporate them into their own vocabulary.

Follow-up activities

- Help your child improve his or her numerical ability by asking them to think of ten things that are fast and ten things that are slow.
- Develop your child's comparative skills by choosing three things from the book and then asking them to list them in order of speed.